Curio & Relics FFL 03 Bound Book
by
Kim Isaac Greenblatt

Kim Greenblatt Publisher
Published In West Hills, California

Curio & Relics FFL 03 Bound Book
by Kim Isaac Greenblatt

Disclosure: You should be aware of the laws and regulations at the Federal and state level where it concerns firearms and the collection of curios and relics. Be safe and do not handle firearms if you are tired, angry, under the influence of alcohol or drugs. Use common sense and take appropriate safety courses. If you find that you are having a problem, an addiction, or can't stop collecting, please seek professional help.

Published by Kim Greenblatt,
West Hills, California, United States of America.

ISBN-13   978-1-60622-004-7
December 2009

Dedicated to people who are collecting. Try not to let it be too all consuming. Your family and friends want to see you once in awhile.

NAME:

STREET ADDRESS:

CITY:

STATE:

ZIP CODE:

PHONE NUMBER:

FFL NUMBER:

STATE LICENSE INFO:

INSURANCE COMPANY INFO:

FIREARMS COLLECTORS ACQUISITION AND DISPOSITION RECORD

Description						Receipt			Disposition				
Manufacturer or Importer	Model	Serial Number	Type	Caliber or Gauge		Date	Name/Add or Name/FFL		Date	Name/Add or Name/FFL	DOB	Drivers Lic or ID	Alien ID

6

FIREARMS COLLECTORS ACQUISITION AND DISPOSITION RECORD

Description					Receipt		Disposition				
Manufacturer or Importer	Model	Serial Number	Type	Caliber or Gauge	Date	Name/Add or Name/FFL	Date	Name/Add or Name/FFL	DOB	Drivers Lic or ID	Alien ID

FIREARMS COLLECTORS ACQUISITION AND DISPOSITION RECORD

Description						Receipt		Disposition			
Manufacturer or Importer	Model	Serial Number	Type	Caliber or Gauge	Date	Name/Add or Name/FFL	Date	Name/Add or Name/FFL	DOB	Drivers Lic or ID	Alien ID

8

FIREARMS COLLECTORS ACQUISITION AND DISPOSITION RECORD

Description						Receipt		Disposition			
Manufacturer or Importer	Model	Serial Number	Type	Caliber or Gauge		Date	Name/Add or Name/FFL	Date	Name/Add or Name/FFL	DOB	Drivers Lic or ID	Alien ID

FIREARMS COLLECTORS ACQUISITION AND DISPOSITION RECORD

Description					Receipt			Disposition				
Manufacturer or Importer	Model	Serial Number	Type	Caliber or Gauge	Date	Name/Add or Name/FFL		Date	Name/Add or Name/FFL	DOB	Drivers Lic or ID	Alien ID

10

FIREARMS COLLECTORS ACQUISITION AND DISPOSITION RECORD

Description					Receipt		Disposition				
Manufacturer or Importer	Model	Serial Number	Type	Caliber or Gauge	Date	Name/Add or Name/FFL	Date	Name/Add or Name/FFL	DOB	Drivers Lic or ID	Alien ID

FIREARMS COLLECTORS ACQUISITION AND DISPOSITION RECORD

Description					Receipt			Disposition			
Manufacturer or Importer	Model	Serial Number	Type	Caliber or Gauge	Date	Name/Add or Name/FFL		Date	Name/Add or Name/FFL	DOB	Drivers Lic or ID	Alien ID

12

FIREARMS COLLECTORS ACQUISITION AND DISPOSITION RECORD

Description						Receipt		Disposition			
Manufacturer or Importer	Model	Serial Number	Type	Caliber or Gauge	Date	Name/Add or Name/FFL	Date	Name/Add or Name/FFL	DOB	Drivers Lic or ID	Alien ID

13

FIREARMS COLLECTORS ACQUISITION AND DISPOSITION RECORD

Description						Receipt		Disposition				
Manufacturer or Importer	Model	Serial Number	Type	Caliber or Gauge		Date	Name/Add or Name/FFL	Date	Name/Add or Name/FFL	DOB	Drivers Lic or ID	Alien ID

FIREARMS COLLECTORS ACQUISITION AND DISPOSITION RECORD

Description					Receipt		Disposition				
Manufacturer or Importer	Model	Serial Number	Type	Caliber or Gauge	Date	Name/Add or Name/FFL	Date	Name/Add or Name/FFL	DOB	Drivers Lic or ID	Alien ID

FIREARMS COLLECTORS ACQUISITION AND DISPOSITION RECORD

Description						Receipt			Disposition				
Manufacturer or Importer	Model	Serial Number	Type	Caliber or Gauge		Date	Name/Add or Name/FFL		Date	Name/Add or Name/FFL	DOB	Drivers Lic or ID	Alien ID

16

FIREARMS COLLECTORS ACQUISITION AND DISPOSITION RECORD

Description					Receipt		Disposition			
Manufacturer or Importer	Model	Serial Number	Type	Caliber or Gauge	Date	Name/Add or Name/FFL	Date	Name/Add or Name/FFL	DOB	Drivers Lic or ID	Alien ID

FIREARMS COLLECTORS ACQUISITION AND DISPOSITION RECORD

Description					Receipt		Disposition			
Manufacturer or Importer	Model	Serial Number	Type	Caliber or Gauge	Date	Name/Add or Name/FFL	Date	Name/Add or Name/FFL	DOB	Drivers Lic or ID	Alien ID

FIREARMS COLLECTORS ACQUISITION AND DISPOSITION RECORD

Description					Receipt		Disposition			
Manufacturer or Importer	Model	Serial Number	Type	Caliber or Gauge	Date	Name/Add or Name/FFL	Date	Name/Add or Name/FFL	DOB	Drivers Lic or ID	Alien ID

19

FIREARMS COLLECTORS ACQUISITION AND DISPOSITION RECORD

Description					Receipt		Disposition				
Manufacturer or Importer	Model	Serial Number	Type	Caliber or Gauge	Date	Name/Add or Name/FFL	Date	Name/Add or Name/FFL	DOB	Drivers Lic or ID	Alien ID

20

FIREARMS COLLECTORS ACQUISITION AND DISPOSITION RECORD

Description					Receipt		Disposition			
Manufacturer or Importer	Model	Serial Number	Type	Caliber or Gauge	Date	Name/Add or Name/FFL	Date	Name/Add or Name/FFL	DOB	Drivers Lic or ID	Alien ID

FIREARMS COLLECTORS ACQUISITION AND DISPOSITION RECORD

Description					Receipt		Disposition			
Manufacturer or Importer	Model	Serial Number	Type	Caliber or Gauge	Date	Name/Add or Name/FFL	Date	Name/Add or Name/FFL	DOB	Drivers Lic or ID	Alien ID

22

FIREARMS COLLECTORS ACQUISITION AND DISPOSITION RECORD

Description					Receipt		Disposition			
Manufacturer or Importer	Model	Serial Number	Type	Caliber or Gauge	Date	Name/Add or Name/FFL	Date	Name/Add or Name/FFL	DOB	Drivers Lic or ID	Alien ID

FIREARMS COLLECTORS ACQUISITION AND DISPOSITION RECORD

Description						Receipt		Disposition			
Manufacturer or Importer	Model	Serial Number	Type	Caliber or Gauge		Date	Name/Add or Name/FFL	Date	Name/Add or Name/FFL	DOB	Drivers Lic or ID	Alien ID

24

FIREARMS COLLECTORS ACQUISITION AND DISPOSITION RECORD

Description					Receipt		Disposition				
Manufacturer or Importer	Model	Serial Number	Type	Caliber or Gauge	Date	Name/Add or Name/FFL	Date	Name/Add or Name/FFL	DOB	Drivers Lic or ID	Alien ID

FIREARMS COLLECTORS ACQUISITION AND DISPOSITION RECORD

Description					Receipt		Disposition			
Manufacturer or Importer	Model	Serial Number	Type	Caliber or Gauge	Date	Name/Add or Name/FFL	Date	Name/Add or Name/FFL	DOB	Drivers Lic or ID	Alien ID

FIREARMS COLLECTORS ACQUISITION AND DISPOSITION RECORD

Description					Receipt		Disposition			
Manufacturer or Importer	Model	Serial Number	Type	Caliber or Gauge	Date	Name/Add or Name/FFL	Date	Name/Add or Name/FFL	DOB	Drivers Lic or ID	Alien ID

FIREARMS COLLECTORS ACQUISITION AND DISPOSITION RECORD

Description					Receipt		Disposition				
Manufacturer or Importer	Model	Serial Number	Type	Caliber or Gauge	Date	Name/Add or Name/FFL	Date	Name/Add or Name/FFL	DOB	Drivers Lic or ID	Alien ID

28

FIREARMS COLLECTORS ACQUISITION AND DISPOSITION RECORD

Description						Receipt		Disposition			
Manufacturer or Importer	Model	Serial Number	Type	Caliber or Gauge		Date	Name/Add or Name/FFL	Date	Name/Add or Name/FFL	DOB	Drivers Lic or ID	Alien ID

FIREARMS COLLECTORS ACQUISITION AND DISPOSITION RECORD

Description					Receipt		Disposition				
Manufacturer or Importer	Model	Serial Number	Type	Caliber of Gauge	Date	Name/Add or Name/FFL	Date	Name/Add or Name/FFL	DOB	Drivers Lic or ID	Alien ID

30

FIREARMS COLLECTORS ACQUISITION AND DISPOSITION RECORD

Description					Receipt			Disposition				
Manufacturer or Importer	Model	Serial Number	Type	Caliber or Gauge	Date	Name/Add or Name/FFL		Date	Name/Add or Name/FFL	DOB	Drivers Lic or ID	Alien ID

FIREARMS COLLECTORS ACQUISITION AND DISPOSITION RECORD

Description					Receipt		Disposition				
Manufacturer or Importer	Model	Serial Number	Type	Caliber or Gauge	Date	Name/Add or Name/FFL	Date	Name/Add or Name/FFL	DOB	Drivers Lic or ID	Alien ID

32

FIREARMS COLLECTORS ACQUISITION AND DISPOSITION RECORD

Description					Receipt			Disposition				
Manufacturer or Importer	Model	Serial Number	Type	Caliber or Gauge	Date	Name/Add or Name/FFL		Date	Name/Add or Name/FFL	DOB	Drivers Lic or ID	Alien ID

FIREARMS COLLECTORS ACQUISITION AND DISPOSITION RECORD

Description					Receipt		Disposition				
Manufacturer or Importer	Model	Serial Number	Type	Caliber or Gauge	Date	Name/Add or Name/FFL	Date	Name/Add or Name/FFL	DOB	Drivers Lic or ID	Alien ID

FIREARMS COLLECTORS ACQUISITION AND DISPOSITION RECORD

Description					Receipt		Disposition			
Manufacturer or Importer	Model	Serial Number	Type	Caliber or Gauge	Date	Name/Add or Name/FFL	Date	Name/Add or Name/FFL	DOB	Drivers Lic or ID	Alien ID

FIREARMS COLLECTORS ACQUISITION AND DISPOSITION RECORD

Description					Receipt		Disposition			
Manufacturer or Importer	Model	Serial Number	Type	Caliber or Gauge	Date	Name/Add or Name/FFL	Date	Name/Add or Name/FFL	DOB	Drivers Lic or ID	Alien ID

FIREARMS COLLECTORS ACQUISITION AND DISPOSITION RECORD

Description						Receipt		Disposition				
Manufacturer or Importer	Model	Serial Number	Type	Caliber or Gauge		Date	Name/Add or Name/FFL	Date	Name/Add or Name/FFL	DOB	Drivers Lic or ID	Alien ID

FIREARMS COLLECTORS ACQUISITION AND DISPOSITION RECORD

Description					Receipt			Disposition				
Manufacturer or Importer	Model	Serial Number	Type	Caliber of Gauge	Date	Name/Add or Name/FFL		Date	Name/Add or Name/FFL	DOB	Drivers Lic or ID	Alien ID

FIREARMS COLLECTORS ACQUISITION AND DISPOSITION RECORD

Description				Receipt		Disposition			
Manufacturer or Importer	Model	Serial Number	Type	Caliber or Gauge	Date	Name/Add or Name/FFL	Date	Name/Add or Name/FFL	DOB	Drivers Lic or ID	Alien ID

FIREARMS COLLECTORS ACQUISITION AND DISPOSITION RECORD

Description						Receipt		Disposition				
Manufacturer or Importer	Model	Serial Number	Type	Caliber or Gauge		Date	Name/Add or Name/FFL	Date	Name/Add or Name/FFL	DOB	Drivers Lic or ID	Alien ID

40

FIREARMS COLLECTORS ACQUISITION AND DISPOSITION RECORD

Description					Receipt		Disposition				
Manufacturer or Importer	Model	Serial Number	Type	Caliber or Gauge	Date	Name/Add or Name/FFL	Date	Name/Add or Name/FFL	DOB	Drivers Lic or ID	Alien ID

FIREARMS COLLECTORS ACQUISITION AND DISPOSITION RECORD

Description					Receipt		Disposition				
Manufacturer or Importer	Model	Serial Number	Type	Caliber or Gauge	Date	Name/Add or Name/FFL	Date	Name/Add or Name/FFL	DOB	Drivers Lic or ID	Alien ID

FIREARMS COLLECTORS ACQUISITION AND DISPOSITION RECORD

Description						Receipt		Disposition			
Manufacturer or Importer	Model	Serial Number	Type	Caliber or Gauge	Date	Name/Add or Name/FFL	Date	Name/Add or Name/FFL	DOB	Drivers Lic or ID	Alien ID

FIREARMS COLLECTORS ACQUISITION AND DISPOSITION RECORD

Description						Receipt		Disposition			
Manufacturer or Importer	Model	Serial Number	Type	Caliber or Gauge		Date	Name/Add or Name/FFL	Date	Name/Add or Name/FFL	DOB	Drivers Lic or ID	Alien ID

FIREARMS COLLECTORS ACQUISITION AND DISPOSITION RECORD

Description					Receipt			Disposition				
Manufacturer or Importer	Model	Serial Number	Type	Caliber or Gauge	Date	Name/Add or Name/FFL	Date	Name/Add or Name/FFL	DOB	Drivers Lic or ID	Alien ID

45

FIREARMS COLLECTORS ACQUISITION AND DISPOSITION RECORD

Description					Receipt		Disposition				
Manufacturer or Importer	Model	Serial Number	Type	Caliber or Gauge	Date	Name/Add or Name/FFL	Date	Name/Add or Name/FFL	DOB	Drivers Lic or ID	Alien ID

FIREARMS COLLECTORS ACQUISITION AND DISPOSITION RECORD

Description						Receipt		Disposition			
Manufacturer or Importer	Model	Serial Number	Type	Caliber or Gauge	Date	Name/Add or Name/FFL	Date	Name/Add or Name/FFL	DOB	Drivers Lic or ID	Alien ID

47

FIREARMS COLLECTORS ACQUISITION AND DISPOSITION RECORD

Description					Receipt		Disposition				
Manufacturer or Importer	Model	Serial Number	Type	Caliber or Gauge	Date	Name/Add or Name/FFL	Date	Name/Add or Name/FFL	DOB	Drivers Lic or ID	Alien ID

FIREARMS COLLECTORS ACQUISITION AND DISPOSITION RECORD

Description					Receipt		Disposition			
Manufacturer or Importer	Model	Serial Number	Type	Caliber or Gauge	Date	Name/Add or Name/FFL	Date	Name/Add or Name/FFL	DOB	Drivers Lic or ID	Alien ID

49

FIREARMS COLLECTORS ACQUISITION AND DISPOSITION RECORD

Description						Receipt		Disposition			
Manufacturer or Importer	Model	Serial Number	Type	Caliber or Gauge		Date	Name/Add or Name/FFL	Date	Name/Add or Name/FFL	DOB	Drivers Lic or ID	Alien ID

50

FIREARMS COLLECTORS ACQUISITION AND DISPOSITION RECORD

Description					Receipt		Disposition				
Manufacturer or Importer	Model	Serial Number	Type	Caliber or Gauge	Date	Name/Add or Name/FFL	Date	Name/Add or Name/FFL	DOB	Drivers Lic or ID	Alien ID

51

FIREARMS COLLECTORS ACQUISITION AND DISPOSITION RECORD

Description					Receipt		Disposition			
Manufacturer or Importer	Model	Serial Number	Type	Caliber or Gauge	Date	Name/Add or Name/FFL	Date	Name/Add or Name/FFL	DOB	Drivers Lic or ID	Alien ID

52

FIREARMS COLLECTORS ACQUISITION AND DISPOSITION RECORD

Description					Receipt			Disposition				
Manufacturer or Importer	Model	Serial Number	Type	Caliber or Gauge	Date	Name/Add or Name/FFL		Date	Name/Add or Name/FFL	DOB	Drivers Lic or ID	Alien ID

FIREARMS COLLECTORS ACQUISITION AND DISPOSITION RECORD

Description					Receipt		Disposition				
Manufacturer or Importer	Model	Serial Number	Type	Caliber or Gauge	Date	Name/Add or Name/FFL	Date	Name/Add or Name/FFL	DOB	Drivers Lic or ID	Alien ID

FIREARMS COLLECTORS ACQUISITION AND DISPOSITION RECORD

Description						Receipt		Disposition			
Manufacturer or Importer	Model	Serial Number	Type	Caliber or Gauge	Date	Name/Add or Name/FFL	Date	Name/Add or Name/FFL	DOB	Drivers Lic or ID	Alien ID

FIREARMS COLLECTORS ACQUISITION AND DISPOSITION RECORD

Description					Receipt		Disposition				
Manufacturer or Importer	Model	Serial Number	Type	Caliber or Gauge	Date	Name/Add or Name/FFL	Date	Name/Add or Name/FFL	DOB	Drivers Lic or ID	Alien ID

FIREARMS COLLECTORS ACQUISITION AND DISPOSITION RECORD

Description					Receipt			Disposition				
Manufacturer or Importer	Model	Serial Number	Type	Caliber or Gauge	Date	Name/Add or Name/FFL		Date	Name/Add or Name/FFL	DOB	Drivers Lic or ID	Alien ID

FIREARMS COLLECTORS ACQUISITION AND DISPOSITION RECORD

Description					Receipt		Disposition				
Manufacturer or Importer	Model	Serial Number	Type	Caliber or Gauge	Date	Name/Add or Name/FFL	Date	Name/Add or Name/FFL	DOB	Drivers Lic or ID	Alien ID

58

FIREARMS COLLECTORS ACQUISITION AND DISPOSITION RECORD

Description						Receipt		Disposition			
Manufacturer or Importer	Model	Serial Number	Type	Caliber or Gauge	Date	Name/Add or Name/FFL	Date	Name/Add or Name/FFL	DOB	Drivers Lic or ID	Alien ID

59

FIREARMS COLLECTORS ACQUISITION AND DISPOSITION RECORD

Description					Receipt		Disposition				
Manufacturer or Importer	Model	Serial Number	Type	Caliber or Gauge	Date	Name/Add or Name/FFL	Date	Name/Add or Name/FFL	DOB	Drivers Lic or ID	Alien ID

FIREARMS COLLECTORS ACQUISITION AND DISPOSITION RECORD

Description					Receipt		Disposition				
Manufacturer or Importer	Model	Serial Number	Type	Caliber or Gauge	Date	Name/Add or Name/FFL	Date	Name/Add or Name/FFL	DOB	Drivers Lic or ID	Alien ID

FIREARMS COLLECTORS ACQUISITION AND DISPOSITION RECORD

Description					Receipt		Disposition			
Manufacturer or Importer	Model	Serial Number	Type	Caliber or Gauge	Date	Name/Add or Name/FFL	Date	Name/Add or Name/FFL	DOB	Drivers Lic or ID	Alien ID

FIREARMS COLLECTORS ACQUISITION AND DISPOSITION RECORD

Description					Receipt		Disposition				
Manufacturer or Importer	Model	Serial Number	Type	Caliber or Gauge	Date	Name/Add or Name/FFL	Date	Name/Add or Name/FFL	DOB	Drivers Lic or ID	Alien ID

63

FIREARMS COLLECTORS ACQUISITION AND DISPOSITION RECORD

Description						Receipt		Disposition			
Manufacturer or Importer	Model	Serial Number	Type	Caliber or Gauge	Date	Name/Add or Name/FFL	Date	Name/Add or Name/FFL	DOB	Drivers Lic or ID	Alien ID

FIREARMS COLLECTORS ACQUISITION AND DISPOSITION RECORD

Description					Receipt		Disposition			
Manufacturer or Importer	Model	Serial Number	Type	Caliber or Gauge	Date	Name/Add or Name/FFL	Date	Name/Add or Name/FFL	DOB	Drivers Lic or ID	Alien ID

FIREARMS COLLECTORS ACQUISITION AND DISPOSITION RECORD

Description						Receipt		Disposition			
Manufacturer or Importer	Model	Serial Number	Type	Caliber or Gauge		Date	Name/Add or Name/FFL	Date	Name/Add or Name/FFL	DOB	Drivers Lic or ID	Alien ID

FIREARMS COLLECTORS ACQUISITION AND DISPOSITION RECORD

Description					Receipt		Disposition				
Manufacturer or Importer	Model	Serial Number	Type	Caliber or Gauge	Date	Name/Add or Name/FFL	Date	Name/Add or Name/FFL	DOB	Drivers Lic or ID	Alien ID

FIREARMS COLLECTORS ACQUISITION AND DISPOSITION RECORD

Description					Receipt		Disposition			
Manufacturer or Importer	Model	Serial Number	Type	Caliber or Gauge	Date	Name/Add or Name/FFL	Date	Name/Add or Name/FFL	DOB	Drivers Lic or ID	Alien ID

68

FIREARMS COLLECTORS ACQUISITION AND DISPOSITION RECORD

Description					Receipt		Disposition				
Manufacturer or Importer	Model	Serial Number	Type	Caliber or Gauge	Date	Name/Add or Name/FFL	Date	Name/Add or Name/FFL	DOB	Drivers Lic or ID	Alien ID

FIREARMS COLLECTORS ACQUISITION AND DISPOSITION RECORD

Description					Receipt		Disposition				
Manufacturer or Importer	Model	Serial Number	Type	Caliber or Gauge	Date	Name/Add or Name/FFL	Date	Name/Add or Name/FFL	DOB	Drivers Lic or ID	Alien ID

70

FIREARMS COLLECTORS ACQUISITION AND DISPOSITION RECORD

Description				Receipt		Disposition			
Manufacturer or Importer	Model	Serial Number	Type	Caliber or Gauge	Date	Name/Add or Name/FFL	Date	Name/Add or Name/FFL	DOB	Drivers Lic or ID	Alien ID

FIREARMS COLLECTORS ACQUISITION AND DISPOSITION RECORD

Description						Receipt			Disposition				
Manufacturer or Importer	Model	Serial Number	Type	Caliber or Gauge		Date	Name/Add or Name/FFL		Date	Name/Add or Name/FFL	DOB	Drivers Lic or ID	Alien ID

72

FIREARMS COLLECTORS ACQUISITION AND DISPOSITION RECORD

Description					Receipt		Disposition			
Manufacturer or Importer	Model	Serial Number	Type	Caliber of Gauge	Date	Name/Add or Name/FFL	Date	Name/Add or Name/FFL	DOB	Drivers Lic or ID	Alien ID

73

FIREARMS COLLECTORS ACQUISITION AND DISPOSITION RECORD

Description					Receipt		Disposition				
Manufacturer or Importer	Model	Serial Number	Type	Caliber or Gauge	Date	Name/Add or Name/FFL	Date	Name/Add or Name/FFL	DOB	Drivers Lic or ID	Alien ID

74

FIREARMS COLLECTORS ACQUISITION AND DISPOSITION RECORD

Description					Receipt		Disposition				
Manufacturer or Importer	Model	Serial Number	Type	Caliber or Gauge	Date	Name/Add or Name/FFL	Date	Name/Add or Name/FFL	DOB	Drivers Lic or ID	Alien ID

FIREARMS COLLECTORS ACQUISITION AND DISPOSITION RECORD

Description					Receipt		Disposition				
Manufacturer or Importer	Model	Serial Number	Type	Caliber or Gauge	Date	Name/Add or Name/FFL	Date	Name/Add or Name/FFL	DOB	Drivers Lic or ID	Alien ID

FIREARMS COLLECTORS ACQUISITION AND DISPOSITION RECORD

Description					Receipt		Disposition				
Manufacturer or Importer	Model	Serial Number	Type	Caliber or Gauge	Date	Name/Add or Name/FFL	Date	Name/Add or Name/FFL	DOB	Drivers Lic or ID	Alien ID

77

FIREARMS COLLECTORS ACQUISITION AND DISPOSITION RECORD

Description					Receipt		Disposition			
Manufacturer or Importer	Model	Serial Number	Type	Caliber or Gauge	Date	Name/Add or Name/FFL	Date	Name/Add or Name/FFL	DOB	Drivers Lic or ID	Alien ID

FIREARMS COLLECTORS ACQUISITION AND DISPOSITION RECORD

Description					Receipt		Disposition			
Manufacturer or Importer	Model	Serial Number	Type	Caliber or Gauge	Date	Name/Add or Name/FFL	Date	Name/Add or Name/FFL	DOB	Drivers Lic or ID	Alien ID

FIREARMS COLLECTORS ACQUISITION AND DISPOSITION RECORD

Description					Receipt			Disposition				
Manufacturer or Importer	Model	Serial Number	Type	Caliber or Gauge	Date	Name/Add or Name/FFL	Date	Name/Add or Name/FFL	DOB	Drivers Lic or ID	Alien ID

FIREARMS COLLECTORS ACQUISITION AND DISPOSITION RECORD

Description					Receipt			Disposition				
Manufacturer or Importer	Model	Serial Number	Type	Caliber or Gauge	Date	Name/Add or Name/FFL		Date	Name/Add or Name/FFL	DOB	Drivers Lic or ID	Alien ID

FIREARMS COLLECTORS ACQUISITION AND DISPOSITION RECORD

Description					Receipt			Disposition				
Manufacturer or Importer	Model	Serial Number	Type	Caliber or Gauge	Date	Name/Add or Name/FFL	Date	Name/Add or Name/FFL	DOB	Drivers Lic or ID	Alien ID	

82

FIREARMS COLLECTORS ACQUISITION AND DISPOSITION RECORD

Description					Receipt		Disposition			
Manufacturer or Importer	Model	Serial Number	Type	Caliber or Gauge	Date	Name/Add or Name/FFL	Date	Name/Add or Name/FFL	DOB	Drivers Lic or ID	Alien ID

FIREARMS COLLECTORS ACQUISITION AND DISPOSITION RECORD

Description					Receipt		Disposition				
Manufacturer or Importer	Model	Serial Number	Type	Caliber or Gauge	Date	Name/Add or Name/FFL	Date	Name/Add or Name/FFL	DOB	Drivers Lic or ID	Alien ID

84

FIREARMS COLLECTORS ACQUISITION AND DISPOSITION RECORD

Description					Receipt			Disposition			
Manufacturer or Importer	Model	Serial Number	Type	Caliber of Gauge	Date	Name/Add or Name/FFL		Date	Name/Add or Name/FFL	DOB	Drivers Lic or ID	Alien ID

FIREARMS COLLECTORS ACQUISITION AND DISPOSITION RECORD

Description					Receipt		Disposition				
Manufacturer or Importer	Model	Serial Number	Type	Caliber or Gauge	Date	Name/Add or Name/FFL	Date	Name/Add or Name/FFL	DOB	Drivers Lic or ID	Alien ID

86

FIREARMS COLLECTORS ACQUISITION AND DISPOSITION RECORD

Description					Receipt		Disposition			
Manufacturer or Importer	Model	Serial Number	Type	Caliber or Gauge	Date	Name/Add or Name/FFL	Date	Name/Add or Name/FFL	DOB	Drivers Lic or ID	Alien ID

FIREARMS COLLECTORS ACQUISITION AND DISPOSITION RECORD

Description					Receipt		Disposition			
Manufacturer or Importer	Model	Serial Number	Type	Caliber or Gauge	Date	Name/Add or Name/FFL	Date	Name/Add or Name/FFL	DOB	Drivers Lic or ID	Alien ID

88

FIREARMS COLLECTORS ACQUISITION AND DISPOSITION RECORD

Description					Receipt		Disposition				
Manufacturer or Importer	Model	Serial Number	Type	Caliber or Gauge	Date	Name/Add or Name/FFL	Date	Name/Add or Name/FFL	DOB	Drivers Lic or ID	Alien ID

FIREARMS COLLECTORS ACQUISITION AND DISPOSITION RECORD

Description					Receipt		Disposition			
Manufacturer or Importer	Model	Serial Number	Type	Caliber or Gauge	Date	Name/Add or Name/FFL	Date	Name/Add or Name/FFL	DOB	Drivers Lic or ID	Alien ID

90

FIREARMS COLLECTORS ACQUISITION AND DISPOSITION RECORD

Description					Receipt		Disposition			
Manufacturer or Importer	Model	Serial Number	Type	Caliber or Gauge	Date	Name/Add or Name/FFL	Date	Name/Add or Name/FFL	DOB	Drivers Lic or ID	Alien ID

FIREARMS COLLECTORS ACQUISITION AND DISPOSITION RECORD

Description						Receipt		Disposition			
Manufacturer or Importer	Model	Serial Number	Type	Caliber or Gauge	Date	Name/Add or Name/FFL	Date	Name/Add or Name/FFL	DOB	Drivers Lic or ID	Alien ID

FIREARMS COLLECTORS ACQUISITION AND DISPOSITION RECORD

Description				Receipt		Disposition			
Manufacturer or Importer	Model	Serial Number	Type	Caliber or Gauge	Date	Name/Add or Name/FFL	Date	Name/Add or Name/FFL	DOB	Drivers Lic or ID	Alien ID

FIREARMS COLLECTORS ACQUISITION AND DISPOSITION RECORD

Description				Receipt			Disposition				
Manufacturer or Importer	Model	Serial Number	Type	Caliber or Gauge	Date	Name/Add or Name/FFL	Date	Name/Add or Name/FFL	DOB	Drivers Lic or ID	Alien ID

FIREARMS COLLECTORS ACQUISITION AND DISPOSITION RECORD

Description					Receipt		Disposition			
Manufacturer or Importer	Model	Serial Number	Type	Caliber or Gauge	Date	Name/Add or Name/FFL	Date	Name/Add or Name/FFL	DOB	Drivers Lic or ID	Alien ID

FIREARMS COLLECTORS ACQUISITION AND DISPOSITION RECORD

Description						Receipt		Disposition			
Manufacturer or Importer	Model	Serial Number	Type	Caliber or Gauge	Date	Name/Add or Name/FFL	Date	Name/Add or Name/FFL	DOB	Drivers Lic or ID	Alien ID

FIREARMS COLLECTORS ACQUISITION AND DISPOSITION RECORD

Description					Receipt			Disposition				
Manufacturer or Importer	Model	Serial Number	Type	Caliber or Gauge	Date	Name/Add or Name/FFL	Date	Name/Add or Name/FFL	DOB	Drivers Lic or ID	Alien ID

FIREARMS COLLECTORS ACQUISITION AND DISPOSITION RECORD

Description					Receipt		Disposition				
Manufacturer or Importer	Model	Serial Number	Type	Caliber or Gauge	Date	Name/Add or Name/FFL	Date	Name/Add or Name/FFL	DOB	Drivers Lic or ID	Alien ID

FIREARMS COLLECTORS ACQUISITION AND DISPOSITION RECORD

Description					Receipt		Disposition			
Manufacturer or Importer	Model	Serial Number	Type	Caliber or Gauge	Date	Name/Add or Name/FFL	Date	Name/Add or Name/FFL	DOB	Drivers Lic or ID	Alien ID

FIREARMS COLLECTORS ACQUISITION AND DISPOSITION RECORD

Description					Receipt		Disposition				
Manufacturer or Importer	Model	Serial Number	Type	Caliber or Gauge	Date	Name/Add or Name/FFL	Date	Name/Add or Name/FFL	DOB	Drivers Lic or ID	Alien ID

FIREARMS COLLECTORS ACQUISITION AND DISPOSITION RECORD

Description					Receipt			Disposition				
Manufacturer or Importer	Model	Serial Number	Type	Caliber or Gauge	Date	Name/Add or Name/FFL		Date	Name/Add or Name/FFL	DOB	Drivers Lic or ID	Alien ID

FIREARMS COLLECTORS ACQUISITION AND DISPOSITION RECORD

Description					Receipt		Disposition				
Manufacturer or Importer	Model	Serial Number	Type	Caliber or Gauge	Date	Name/Add or Name/FFL	Date	Name/Add or Name/FFL	DOB	Drivers Lic or ID	Alien ID

102

FIREARMS COLLECTORS ACQUISITION AND DISPOSITION RECORD

Description					Receipt		Disposition			
Manufacturer or Importer	Model	Serial Number	Type	Caliber or Gauge	Date	Name/Add or Name/FFL	Date	Name/Add or Name/FFL	DOB	Drivers Lic or ID	Alien ID

FIREARMS COLLECTORS ACQUISITION AND DISPOSITION RECORD

Description					Receipt		Disposition			
Manufacturer or Importer	Model	Serial Number	Type	Caliber or Gauge	Date	Name/Add or Name/FFL	Date	Name/Add or Name/FFL	DOB	Drivers Lic or ID	Alien ID

FIREARMS COLLECTORS ACQUISITION AND DISPOSITION RECORD

Description					Receipt		Disposition			
Manufacturer or Importer	Model	Serial Number	Type	Caliber or Gauge	Date	Name/Add or Name/FFL	Date	Name/Add or Name/FFL	DOB	Drivers Lic or ID	Alien ID

FIREARMS COLLECTORS ACQUISITION AND DISPOSITION RECORD

Description					Receipt		Disposition			
Manufacturer or Importer	Model	Serial Number	Type	Caliber or Gauge	Date	Name/Add or Name/FFL	Date	Name/Add or Name/FFL	DOB	Drivers Lic or ID	Alien ID

FIREARMS COLLECTORS ACQUISITION AND DISPOSITION RECORD

Description					Receipt		Disposition			
Manufacturer or Importer	Model	Serial Number	Type	Caliber or Gauge	Date	Name/Add or Name/FFL	Date	Name/Add or Name/FFL	DOB	Drivers Lic or ID	Alien ID

FIREARMS COLLECTORS ACQUISITION AND DISPOSITION RECORD

Description				**Receipt**			**Disposition**				
Manufacturer or Importer	Model	Serial Number	Type	Caliber or Gauge	**Date**	Name/Add or Name/FFL	**Date**	Name/Add or Name/FFL	DOB	Drivers Lic or ID	Alien ID

108

FIREARMS COLLECTORS ACQUISITION AND DISPOSITION RECORD

Description					Receipt		Disposition				
Manufacturer or Importer	Model	Serial Number	Type	Caliber or Gauge	Date	Name/Add or Name/FFL	Date	Name/Add or Name/FFL	DOB	Drivers Lic or ID	Alien ID

FIREARMS COLLECTORS ACQUISITION AND DISPOSITION RECORD

Description					Receipt		Disposition			
Manufacturer or Importer	Model	Serial Number	Type	Caliber or Gauge	Date	Name/Add or Name/FFL	Date	Name/Add or Name/FFL	DOB	Drivers Lic or ID	Alien ID

110

FIREARMS COLLECTORS ACQUISITION AND DISPOSITION RECORD

Description					Receipt		Disposition			
Manufacturer or Importer	Model	Serial Number	Type	Caliber or Gauge	Date	Name/Add or Name/FFL	Date	Name/Add or Name/FFL	DOB	Drivers Lic or ID	Alien ID

111

FIREARMS COLLECTORS ACQUISITION AND DISPOSITION RECORD

Description					Receipt			Disposition				
Manufacturer or Importer	Model	Serial Number	Type	Caliber or Gauge	Date	Name/Add or Name/FFL	Date	Name/Add or Name/FFL	DOB	Drivers Lic or ID	Alien ID	

112

FIREARMS COLLECTORS ACQUISITION AND DISPOSITION RECORD

Description					Receipt		Disposition			
Manufacturer or Importer	Model	Serial Number	Type	Caliber or Gauge	Date	Name/Add or Name/FFL	Date	Name/Add or Name/FFL	DOB	Drivers Lic or ID	Alien ID

FIREARMS COLLECTORS ACQUISITION AND DISPOSITION RECORD

Description					Receipt		Disposition				
Manufacturer or Importer	Model	Serial Number	Type	Caliber or Gauge	Date	Name/Add or Name/FFL	Date	Name/Add or Name/FFL	DOB	Drivers Lic or ID	Alien ID

114

FIREARMS COLLECTORS ACQUISITION AND DISPOSITION RECORD

Description						Receipt		Disposition			
Manufacturer or Importer	Model	Serial Number	Type	Caliber or Gauge		Date	Name/Add or Name/FFL	Date	Name/Add or Name/FFL	DOB	Drivers Lic or ID	Alien ID

FIREARMS COLLECTORS ACQUISITION AND DISPOSITION RECORD

Description					Receipt		Disposition				
Manufacturer or Importer	Model	Serial Number	Type	Caliber or Gauge	Date	Name/Add or Name/FFL	Date	Name/Add or Name/FFL	DOB	Drivers Lic or ID	Alien ID

116

FIREARMS COLLECTORS ACQUISITION AND DISPOSITION RECORD

Description					Receipt		Disposition				
Manufacturer or Importer	Model	Serial Number	Type	Caliber or Gauge	Date	Name/Add or Name/FFL	Date	Name/Add or Name/FFL	DOB	Drivers Lic or ID	Alien ID

117

FIREARMS COLLECTORS ACQUISITION AND DISPOSITION RECORD

Description					Receipt		Disposition			
Manufacturer or Importer	Model	Serial Number	Type	Caliber or Gauge	Date	Name/Add or Name/FFL	Date	Name/Add or Name/FFL	DOB	Drivers Lic or ID	Alien ID